Chemistry Made Easy: A Student's Guide to Understanding the Elements

Mark

Copyright © [2023]

Author: Nadia

Title: Chemistry Made Easy: A Student's Guide to Understanding the Elements

This book is a product of [Publisher's Mark]

ISBN:

TABLE OF CONTENTS

Chapter 15: Conclusion and Further Resources
99

Recap of Key Points

Additional Resources for Further Study

Final Thoughts and Encouragement

Chapter 1: Introduction to Chemistry

The Importance of Chemistry

Chemistry is often perceived as a complex and daunting subject, but understanding its importance is crucial for students in the realm of science education. In this subchapter, we will explore the significance of chemistry and how it impacts various aspects of our lives.

To begin with, chemistry is the foundation of all matter. It unravels the mysteries behind the elements, compounds, and molecules that make up our world. By studying chemistry, students gain a better understanding of the fundamental building blocks of nature, enabling them to comprehend the physical and chemical properties of substances.

Moreover, chemistry plays a pivotal role in everyday life. From the food we eat to the medicines we take, chemistry is at the heart of it all. For instance, understanding the chemical reactions involved in cooking allows us to create delicious meals by manipulating ingredients and applying heat appropriately. Similarly, pharmaceutical advancements rely heavily on the knowledge of chemistry, enabling the development of life-saving drugs that combat diseases and improve human health.

Furthermore, chemistry provides a deeper understanding of the environment and the challenges we face today. As students, learning about the chemical reactions involved in global warming, air pollution, and water contamination helps us comprehend the impact of human activities on our planet. Armed with this knowledge, we can develop innovative solutions and contribute to the preservation of our environment.

Chemistry also plays a significant role in technological advancements. From the development of new materials to the creation of sustainable energy sources, chemistry is the driving force behind innovation. By studying chemistry, students gain the skills necessary to contribute to breakthroughs in fields such as nanotechnology, renewable energy, and medical research.

In conclusion, the importance of chemistry cannot be overstated. It is the key to understanding the world around us, from the smallest particle to the vastness of the universe. By delving into the realm of chemistry, students gain a solid foundation in science education, enabling them to contribute to advancements in various fields. Whether it be improving our health, preserving the environment, or driving technological innovation, the knowledge of chemistry empowers students to make a positive impact on society.

Overview of the Elements

In the vast world of chemistry, the elements are the building blocks that make up everything around us. From the air we breathe to the water we drink, everything is composed of these fundamental substances. Understanding the elements is crucial in unraveling the mysteries of the universe and gaining a deeper knowledge of the world we live in. In this subchapter, we will delve into the fascinating world of elements and explore their properties and significance.

The periodic table is a comprehensive chart that organizes all the known elements. It is a powerful tool that allows scientists and students alike to understand and categorize the elements based on their atomic number, atomic mass, and chemical properties. From the noble gases to the transition metals, each element has its own unique characteristics and plays a vital role in different chemical reactions.

Elements are classified into various groups and periods based on their properties. Understanding these classifications helps us identify patterns and trends in the behavior of elements. For example, elements in the same group often exhibit similar chemical properties, while elements in the same period have the same number of electron shells.

The subchapter will explore some of the most important elements and their significance in various fields of science. We will delve into the properties of elements such as hydrogen, oxygen, carbon, nitrogen, and many others. These elements are not only essential for life on Earth but also have diverse applications in industry, medicine, and technology.

Moreover, we will discuss the concept of atomic structure and how it determines the behavior of elements. Understanding the arrangement of protons, neutrons, and electrons in an atom is crucial in comprehending the reactivity and stability of elements.

Throughout the subchapter, we will use real-life examples and practical applications to make the study of elements relatable and engaging for students. From the uses of oxygen in fire to the role of carbon in organic compounds, we will connect the dots between the elements and their impact on our daily lives.

By the end of this subchapter, students will have a solid foundation in understanding the elements and their significance in the world of chemistry. Armed with this knowledge, they will be better equipped to explore the fascinating world of chemical reactions, compounds, and materials.

In "Chemistry Made Easy: A Student's Guide to Understanding the Elements," this subchapter is designed to provide a comprehensive overview of the elements, catering specifically to students in the field of science education.

Understanding Chemical Reactions

Chemistry is the study of matter and its transformations. One of the key concepts in chemistry is chemical reactions, which involve the rearrangement of atoms to form new substances. Understanding how chemical reactions occur is crucial for students of chemistry as it provides a foundation for understanding the behavior of various elements and compounds.

Chemical reactions occur when two or more substances, called reactants, interact with each other and transform into different substances, known as products. These reactions are governed by the laws of conservation of mass and energy. In other words, the total mass and energy before and after a reaction remains the same.

To understand chemical reactions, it is important to grasp the concept of chemical equations. Chemical equations are symbolic representations of reactions, which show the reactants on the left side and the products on the right side, separated by an arrow. For example, the equation for the reaction between hydrogen gas and oxygen gas to form water is:

$$2H_2 + O_2 \rightarrow 2H_2O$$

In this equation, the numbers in front of the chemical formulas are called coefficients and indicate the relative amounts of each substance involved in the reaction.

Chemical reactions can be classified into different types based on their characteristics. Some common types include synthesis reactions, decomposition reactions, single displacement reactions, double displacement reactions, and combustion reactions. Each type of reaction follows specific patterns and has unique properties.

Understanding chemical reactions also involves recognizing the factors that influence the rate of reactions. These factors include temperature, concentration, surface area, and the presence of catalysts. By manipulating these factors, scientists can control the speed at which a reaction occurs.

Chemical reactions play a vital role in our everyday lives. From the food we eat to the fuels we burn, everything around us involves chemical reactions. By understanding the principles behind chemical reactions, students can not only appreciate the world around them but also make informed decisions regarding their health, environment, and technological advancements.

In conclusion, understanding chemical reactions is essential for students studying chemistry. By comprehending the basics of chemical equations, types of reactions, and factors that influence reaction rates, students can gain a deeper understanding of the behavior of elements and compounds. Chemical reactions are not only fundamental to chemistry but also have real-world applications that impact our lives.

Chapter 2: The Periodic Table

History and Development of the Periodic Table

The Periodic Table is the cornerstone of chemistry, providing a systematic way to organize and understand the elements. But have you ever wondered how this incredible tool came to be? In this subchapter, we will take a journey through the history and development of the Periodic Table, unraveling the fascinating story behind its creation.

The story begins in the late 18th century when scientists began to explore the properties of various elements. They observed that certain elements shared similar characteristics and tried to find a way to organize them. However, it wasn't until the early 19th century that the first attempts at creating a rudimentary version of the Periodic Table were made.

One of the early pioneers was Johann Wolfgang Döbereiner, who noticed that some elements could be grouped into triads based on their properties. For example, chlorine, bromine, and iodine shared similar chemical behavior. This idea of grouping elements based on their properties laid the groundwork for future developments.

In the mid-19th century, a breakthrough came from the work of Dmitri Mendeleev. He organized the known elements according to their atomic weights and noticed a repeating pattern of properties. Mendeleev's Periodic Table not only arranged the elements in order of increasing atomic weight but also left gaps for undiscovered elements, predicting their properties accurately.

As the years went by, other scientists contributed to the development of the Periodic Table. In the early 20th century, Henry Moseley discovered that the elements' properties were better correlated with their atomic numbers rather than their atomic weights. This led to the modern understanding of the Periodic Table, where elements are arranged in order of increasing atomic number.

The Periodic Table we know today is a testament to the collective effort of countless scientists who dedicated their lives to understanding the elements. It has gone through several revisions and improvements, incorporating new elements and refining our knowledge of chemical behavior.

Understanding the history and development of the Periodic Table is crucial for any aspiring chemist. It not only helps us appreciate the ingenuity of the scientists who came before us but also provides a solid foundation for understanding the patterns and trends that govern the behavior of elements.

In conclusion, the Periodic Table is a remarkable achievement that revolutionized the field of chemistry. Its development, starting from the observation of similar properties to the modern understanding of atomic numbers, represents the tireless efforts of scientists throughout history. By exploring this fascinating journey, we can gain a deeper appreciation for the elements that make up our world and the incredible tool that allows us to understand them – the Periodic Table.

Organization of the Periodic Table

The periodic table is an essential tool in the study of chemistry. It is a comprehensive arrangement of all the known elements, providing a systematic way to understand their properties and relationships. In this subchapter, we will delve into the organization of the periodic table, exploring its structure and significance.

At first glance, the periodic table may seem like a jumble of numbers and symbols. However, it is carefully organized based on the properties of elements. The table is divided into rows called periods and columns called groups. Each element is represented by a unique symbol and arranged in increasing order of atomic number, which corresponds to the number of protons in an atom's nucleus.

One of the most notable features of the periodic table is the presence of periodic trends. These trends refer to regular patterns observed in the properties of elements as we move across a period or down a group. For instance, as we progress from left to right across a period, the atomic radius generally decreases, while the ionization energy increases. These trends provide valuable insights into the behavior of elements and help us predict their chemical properties.

The periodic table also highlights the distinction between metals, nonmetals, and metalloids. Metals, located on the left side, exhibit characteristics such as high thermal and electrical conductivity, malleability, and ductility. Nonmetals, on the right side, tend to be poor conductors of heat and electricity and are usually brittle. Metalloids, found along the staircase dividing metals and nonmetals, possess some properties of both groups.

Furthermore, the periodic table allows us to identify elements with similar chemical properties. Elements within the same group often exhibit similar valence electron configurations, leading to comparable reactivity. This grouping aids in classifying elements into families, such as alkali metals, alkaline earth metals, halogens, and noble gases.

Understanding the organization of the periodic table is crucial for several reasons. Firstly, it simplifies the study of chemistry by providing a structured framework to navigate the vast array of elements. Secondly, it enables us to make predictions about the properties and behavior of unknown elements. Lastly, it serves as a vital reference tool for scientists, allowing them to access information about elements efficiently.

In conclusion, the organization of the periodic table is a fundamental aspect of chemistry education. By comprehending its structure and significance, students can develop a solid foundation in understanding the elements. The periodic table's systematic arrangement, periodic trends, and grouping of elements based on properties provide invaluable insights into the world of chemistry and facilitate further discoveries in the field.

Properties and Characteristics of Elements

In the fascinating realm of chemistry, one of the fundamental concepts is the study of elements. Elements are the building blocks of matter, and understanding their properties and characteristics is crucial to comprehending the intricate workings of the world around us. This subchapter aims to provide students with a comprehensive overview of the properties and characteristics of elements, offering a solid foundation for further exploration in the field of chemistry.

At the heart of the study of elements lies the periodic table. This powerful tool organizes elements based on their atomic number, electron configuration, and recurring chemical properties. By understanding the periodic table, students gain insight into the characteristics of each element, such as its atomic mass, symbol, and atomic structure.

One vital property of elements is their atomic structure. Elements are defined by the number of protons in their nucleus, known as the atomic number. This number determines an element's position in the periodic table and provides information about its chemical behavior. Additionally, the arrangement of electrons around the nucleus plays a crucial role in determining an element's reactivity and bonding capabilities.

Another key characteristic of elements is their chemical reactivity. Some elements, located on the left side of the periodic table, are highly reactive and readily form compounds with other elements. These elements, known as metals, exhibit properties such as malleability, conductivity, and luster. On the other hand, elements located on the right side of the periodic table, called nonmetals, tend to be less reactive and often form compounds by gaining electrons or sharing them with other elements.

The periodic table also reveals periodic trends, which are patterns in the properties of elements. For example, as one moves from left to right across a period, elements generally become less metallic and more nonmetallic. Similarly, as one moves down a group, elements tend to increase in size and reactivity.

Furthermore, elements possess various physical properties that define their behavior under different conditions. These properties include boiling point, melting point, density, and color, among others. Understanding these properties allows scientists to predict how elements will interact with one another and how they can be utilized in various applications.

In summary, the properties and characteristics of elements are essential knowledge for any student of chemistry. By comprehending the periodic table, atomic structure, chemical reactivity, periodic trends, and physical properties, students gain a firm grasp of the foundations of chemistry. With this knowledge, they can explore the fascinating world of elements, unlocking the mysteries of matter and paving the way for further scientific discoveries.

Chapter 3: Atomic Structure

The Atom: Fundamental Building Block

In the vast realm of chemistry, understanding the atom is the key to unlocking the secrets of the elements. At its core, the atom is the fundamental building block of matter, and without a grasp of its structure and behavior, comprehending the intricacies of chemical reactions becomes an insurmountable challenge. This subchapter will delve into the captivating world of atoms, providing a comprehensive overview that will leave students with a solid foundation in this essential concept.

To begin our exploration, it is crucial to grasp the concept that all matter is composed of atoms. These tiny particles are incredibly small, so much so that a single grain of sand contains more atoms than there are stars in the observable universe. The atom consists of a central nucleus, which contains positively charged protons and neutral neutrons, while negatively charged electrons orbit around the nucleus in specific energy levels or shells.

The atomic structure determines an element's unique properties, such as its atomic number, mass number, and chemical reactivity. The atomic number represents the number of protons in an atom's nucleus, while the mass number is the sum of protons and neutrons. Students will learn how to read the periodic table, which categorizes elements based on their atomic number and provides an abundance of information about each element.

Moreover, this subchapter will introduce the concept of isotopes, which are atoms of the same element with different numbers of neutrons. Isotopes can have varying masses but exhibit similar chemical behavior. Students will discover how isotopes are used in various fields, such as radiocarbon dating and medical imaging.

In addition to the atom's structure, students will explore the behavior of electrons in atoms. They will learn about energy levels and how electrons occupy specific orbitals within these levels. The concept of valence electrons will be discussed, as they play a significant role in determining an element's reactivity and its ability to form bonds.

Understanding the atom is the cornerstone of chemistry. From the simplest compound to the most complex reactions, the behavior of atoms governs it all. By grasping the atom's structure, students will gain a deep appreciation for the complexity and beauty of the elements. Through engaging explanations, relatable examples, and interactive activities, this subchapter aims to make the atom come alive for students, providing them with a solid foundation to build upon as they delve further into the fascinating world of chemistry.

Subatomic Particles: Protons, Neutrons, and Electrons

In the fascinating world of chemistry, understanding the building blocks of matter is crucial. At the heart of every atom lie three essential subatomic particles: protons, neutrons, and electrons. Together, they form the foundation of the elements and dictate their unique properties.

Protons, as the name suggests, carry a positive charge. These tiny particles are found within the nucleus of an atom. With a relative mass of 1, protons play a pivotal role in determining an element's identity. For example, an atom with one proton is hydrogen, while an atom with six protons is carbon. The number of protons in an atom is known as the atomic number, which defines its place on the periodic table.

Neutrons, on the other hand, are electrically neutral particles found alongside protons in the nucleus. With a relative mass similar to protons, neutrons help stabilize the nucleus and prevent it from breaking apart. The different isotopes of an element arise from variations in the number of neutrons present in the nucleus. These isotopes may have slightly different properties, such as stability or radioactivity.

Finally, electrons are negatively charged particles that orbit the nucleus in specific energy levels called shells or orbitals. Unlike protons and neutrons, electrons have a negligible mass. It is the arrangement and movement of electrons that determine an atom's chemical behavior. Electrons occupy specific energy levels, with the lowest energy level closest to the nucleus. Each shell can hold a limited number of electrons, with the first shell holding a maximum of two, the second shell up to eight, and so on.

Understanding the roles and interactions of these subatomic particles is vital in comprehending the behavior of atoms and the elements they form. The arrangement of protons, neutrons, and electrons within an atom determines its atomic mass, charge, and reactivity. By gaining knowledge of these particles, students can unlock the secrets of the periodic table and delve deeper into the world of chemistry.

In summary, protons, neutrons, and electrons are the three fundamental subatomic particles that make up atoms. Protons define an element's identity, neutrons stabilize the nucleus, and electrons determine an atom's chemical behavior. By studying these particles, students can unravel the mysteries of the elements and develop a solid foundation in the captivating world of chemistry.

Atomic Number and Mass Number

In the exciting world of chemistry, it is crucial to understand the fundamental concepts that underpin the behavior of the elements. Two such concepts are the atomic number and mass number, which provide essential information about an element's identity and characteristics. In this subchapter, we will delve into these concepts, demystifying them and empowering you with a deeper understanding of the elements.

The atomic number of an element refers to the number of protons found within the nucleus of its atoms. It acts as a unique identifier for each element, as no two elements have the same number of protons. For instance, hydrogen, the first element on the periodic table, has an atomic number of 1, signifying that its atoms contain a single proton. Oxygen, on the other hand, has an atomic number of 8, indicating the presence of eight protons in its atoms. As you progress through the periodic table, you will notice that the atomic number increases incrementally, providing a systematic arrangement of elements.

While the atomic number is specific to each element, the mass number provides information about the total number of particles within an atom's nucleus. It is defined as the sum of protons and neutrons present in the nucleus. Neutrons, unlike protons, do not contribute to the element's identity but play a crucial role in determining its isotopes. Isotopes are atoms of the same element with different numbers of neutrons. For instance, carbon-12 and carbon-14 are isotopes of carbon, differing in their mass numbers due to variations in neutron counts.

Understanding the atomic number and mass number allows us to determine other important properties of elements, such as their atomic mass and charge. The atomic mass is the weighted average of the masses of an element's isotopes, taking into account the relative abundance of each isotope in nature. It can be calculated using the mass number and the percentage abundance of each isotope. The atomic charge, or number of electrons, is equal to the number of protons in a neutral atom. It plays a crucial role in chemical reactions and the formation of compounds.

In conclusion, the atomic number and mass number are vital concepts in chemistry that provide key information about an element's identity and characteristics. By understanding these concepts, you will be able to navigate the periodic table with ease and comprehend the behavior of elements in various chemical reactions. So, let's continue our exploration of the fascinating world of elements and unlock the secrets they hold.

Chapter 4: Chemical Bonds

Types of Chemical Bonds

In the world of chemistry, the interaction between atoms is crucial for the formation of compounds and molecules. These interactions are known as chemical bonds and they play a fundamental role in determining the properties and behavior of different substances. Understanding the types of chemical bonds is essential for any student of chemistry. Let's explore the three primary types of chemical bonds: ionic bonds, covalent bonds, and metallic bonds.

Ionic bonds occur when there is a transfer of electrons between atoms. This type of bond typically forms between a metal and a non-metal. In an ionic bond, one atom loses electrons (becoming a positively charged ion) and another atom gains those electrons (becoming a negatively charged ion). The resulting attraction between the oppositely charged ions gives rise to the ionic bond. Ionic compounds are usually solid and have high melting and boiling points, such as sodium chloride (table salt).

Covalent bonds, on the other hand, involve the sharing of electrons between atoms. This type of bond typically forms between non-metals. In a covalent bond, the atoms share one or more pairs of electrons to achieve a stable electron configuration. Covalent compounds can exist as solids, liquids, or gases, and they often have lower melting and boiling points compared to ionic compounds. Examples of covalent compounds include water (H_2O) and methane (CH_4).

Lastly, metallic bonds occur between metal atoms. In a metallic bond, the valence electrons of the atoms are delocalized, meaning they are free to move throughout the entire metal lattice. This shared "sea" of electrons gives metals their unique properties, such as thermal and electrical conductivity, malleability, and ductility. Metallic bonds are responsible for the shiny appearance and high density of metals, as well as their ability to conduct heat and electricity.

Understanding the different types of chemical bonds is crucial for predicting and explaining the behavior and properties of substances. It provides insights into the interactions between atoms and how they contribute to the formation of compounds. By grasping the concepts of ionic, covalent, and metallic bonds, students can build a solid foundation in chemistry and apply this knowledge to various scientific fields.

In summary, chemical bonds are the forces that hold atoms together in compounds and molecules. Ionic bonds involve the transfer of electrons, covalent bonds involve the sharing of electrons, and metallic bonds involve a sea of delocalized electrons. Each type of bond has its unique characteristics and properties, which are essential for understanding the behavior of substances. By exploring and comprehending these different types of chemical bonds, students can unlock the secrets of the elements and delve deeper into the fascinating world of chemistry.

Ionic Bonds: Formation and Properties

In the vast realm of chemistry, few topics are as intriguing and fundamental as ionic bonds. Understanding how these bonds form and their unique properties is crucial for any aspiring scientist. In this subchapter, we will delve into the fascinating world of ionic bonds, offering a comprehensive explanation tailored to students who are passionate about science education.

Ionic bonds are the result of the attractive forces between positively and negatively charged ions. These ions are formed when atoms gain or lose electrons, thereby acquiring a charge. Atoms with a surplus of electrons become negatively charged ions, known as anions, while atoms with a deficit of electrons become positively charged ions, known as cations.

The formation of ionic bonds occurs when these oppositely charged ions are brought together. The cations and anions are drawn to each other by electrostatic forces, resulting in a strong bond. This bond is typically formed between a metal and a non-metal, as metals tend to lose electrons, while non-metals tend to gain them.

One of the defining properties of ionic bonds is their ability to form crystal lattices. Due to the strong electrostatic attraction between the ions, they arrange themselves in a repeating pattern, creating a three-dimensional structure. This regular arrangement gives ionic compounds their characteristic crystalline appearance.

Ionic compounds also exhibit high melting and boiling points. The strong ionic bonds require a significant amount of energy to break, which is why these compounds often exist as solids at room temperature. When heated, the lattice structure gradually collapses, leading to the compound's transition from a solid to a liquid or gas state.

Another noteworthy property of ionic compounds is their tendency to conduct electricity when dissolved in water or melted. In their solid state, ions are locked in place and cannot move. However, when dissolved or melted, the ions become mobile and can carry electric charge, allowing the solution or molten compound to conduct electricity.

Understanding the formation and properties of ionic bonds is essential not only for grasping the intricacies of chemistry but also for comprehending the nature of matter itself. The study of ionic bonds paves the way for comprehending more complex chemical processes and reactions.

In conclusion, this subchapter has provided a comprehensive overview of ionic bonds, their formation, and properties. By understanding the intricate dance between positively and negatively charged ions, students can unlock the secrets of chemical bonding and lay a solid foundation for further exploration in the captivating world of chemistry.

Covalent Bonds: Sharing of Electrons

In the vast world of chemistry, one of the fundamental concepts that every student needs to grasp is the concept of covalent bonds. Covalent bonds are all about the sharing of electrons between atoms, and understanding this concept is crucial in comprehending the behavior of elements and compounds.

So, what exactly is a covalent bond? Well, it occurs when two atoms share electrons in order to achieve a stable configuration. You can think of it as a partnership where both atoms contribute and benefit from the shared electrons. This sharing of electrons allows both atoms to fill their outermost electron shells, resulting in a more stable and balanced arrangement.

Let's take an example to illustrate this concept further. Consider a molecule of water (H_2O). In this case, two hydrogen atoms and one oxygen atom form a covalent bond. The oxygen atom, which has six valence electrons, shares two of its electrons with each hydrogen atom. In return, the hydrogen atoms, which have only one valence electron, share their electrons with the oxygen atom. This sharing of electrons creates a stable water molecule.

It's important to note that not all covalent bonds are the same. They can vary in strength and polarity. The strength of a covalent bond depends on the number of shared electrons and the distance between the atoms. The closer the atoms are, the stronger the bond.

Additionally, covalent bonds can be categorized as polar or nonpolar. In a polar covalent bond, the electrons are not shared equally between the atoms, resulting in a partial positive charge on one atom and a partial negative charge on the other. On the other hand, in a nonpolar covalent bond, the electrons are shared equally, resulting in no charge separation.

Understanding covalent bonds is crucial for understanding the properties and behavior of molecules. It helps explain why certain substances have specific melting and boiling points, why some compounds dissolve in water while others don't, and even why certain molecules have unique shapes and structures.

In conclusion, covalent bonds play a vital role in the world of chemistry. By sharing electrons, atoms are able to achieve stability and form molecules with unique properties. Understanding the concept of covalent bonds is essential for any student delving into the fascinating realm of chemistry and scientific exploration.

Metallic Bonds: The Bonding in Metals

In the world of Chemistry, understanding the different types of chemical bonds is crucial to comprehending the behavior of elements and compounds. One such type of bond is the metallic bond, which plays a significant role in the properties and characteristics of metals. In this subchapter, we will delve into the fascinating world of metallic bonds and explore how they contribute to the unique properties of metals.

At its core, a metallic bond is a type of chemical bond that occurs between metal atoms. Unlike ionic or covalent bonds, metallic bonds involve a sharing of electrons between all the metal atoms in a substance. This sharing creates a pool of electrons, often referred to as a "sea of electrons," that is free to move throughout the entire metal lattice. This delocalized electron cloud is what gives metals their distinctive properties.

One of the key features of metallic bonds is their strength. The sharing of electrons amongst a large number of metal atoms results in a strong bond, making metals generally solid at room temperature. This is why metals are known for their high melting and boiling points, as well as their ability to withstand considerable pressure.

Another important characteristic of metallic bonds is their ability to conduct electricity and heat. The delocalized electrons in the metal lattice can move freely, allowing for the efficient transfer of both thermal energy and electric current. This is why metals are excellent conductors of electricity and are commonly used in electrical wiring and circuits.

Furthermore, metallic bonds also contribute to the malleability and ductility of metals. The free movement of electrons enables metal atoms to slide past each other without breaking the bond. This property allows metals to be hammered into thin sheets (malleability) and drawn into wires (ductility) without losing their structural integrity.

Understanding metallic bonds is vital not only for understanding the properties of metals but also for various applications in fields such as engineering, materials science, and even pharmacology. It is through the manipulation and understanding of metallic bonds that scientists and engineers can develop new and improved metal alloys for a wide range of applications.

In conclusion, metallic bonds are a unique type of chemical bond that occurs between metal atoms. The sharing of electrons creates a strong bond and a sea of delocalized electrons that contribute to the exceptional properties of metals. From their high melting points and conductivity to their malleability and ductility, metallic bonds are fundamental to our understanding of metals. By comprehending the nature of metallic bonds, students can gain a deeper appreciation for the world of Chemistry and its practical applications in various scientific and technological fields.

Chapter 5: States of Matter

Solid, Liquid, and Gas: The Three States of Matter

In the world of chemistry, matter is classified into three main states: solid, liquid, and gas. Understanding these states is crucial in comprehending the behavior and properties of various substances. So, let's delve into the fascinating world of matter and explore the characteristics of each state.

Solid: The first state of matter we'll explore is solid. Solids have a definite shape and volume, meaning they maintain a rigid structure and cannot be compressed easily. Picture a block of ice – it retains its shape unless an external force acts upon it. The particles in a solid are closely packed together, which explains their stability. Examples of solids include wood, metals, rocks, and even your textbooks.

Liquid: Moving on to the second state of matter, liquids are substances that have a definite volume but take the shape of their container. Unlike solids, liquids can flow and are not rigid. When you pour water into a glass, it takes the shape of the glass. Liquids can also be compressed slightly, but not to the extent of gases. The particles in liquids are more loosely packed than in solids, allowing them to move past one another. Common examples of liquids are water, milk, oil, and fruit juices.

Gas: Lastly, we have the gaseous state of matter. Gases have neither a definite shape nor volume. They expand to fill the entire space available to them, regardless of the size or shape of the container. Think of the air around you – it occupies the entire room you are in.

Gases are highly compressible due to the large spaces between particles. The particles in gases have the most freedom of movement, constantly colliding with one another. Examples of gases include oxygen, nitrogen, carbon dioxide, and helium.

Understanding the different states of matter is essential because substances can undergo changes in their state under certain conditions. For instance, heating a solid can transform it into a liquid, and further heating can convert it into a gas. This knowledge is not only fascinating but also helps us make sense of the world around us.

So, as you continue your journey through the fascinating realm of chemistry, remember that matter can exist in three distinct states: solid, liquid, and gas. Each state has its own unique properties, and understanding these properties will enable you to comprehend the behavior of substances in different situations. Keep exploring and uncovering the mysteries of matter – the building blocks of our universe!

Changes in States of Matter

Understanding the changes that occur in the states of matter is essential in the study of chemistry. In this subchapter, we will explore the various transformations that take place between solids, liquids, and gases. By understanding these changes, students will gain a deeper appreciation for the behavior and properties of different substances.

States of matter refer to the physical forms in which matter can exist: solid, liquid, and gas. Each state has unique characteristics that set it apart from the others. Solids have a definite shape and volume, while liquids take the shape of their containers but have a definite volume. Gases, on the other hand, have neither a definite shape nor volume, as they expand to fill any container they occupy.

One of the fundamental concepts in this subchapter is the idea of phase changes. Phase changes occur when a substance undergoes a transformation from one state to another. These changes are driven by alterations in temperature and pressure.

The most common phase changes include melting, freezing, evaporation, condensation, sublimation, and deposition. Melting is the process by which a solid becomes a liquid, while freezing is the reverse transformation. Evaporation occurs when a liquid turns into a gas, and condensation is the opposite process. Sublimation refers to the direct transformation of a solid into a gas, and deposition is the reverse process.

Understanding the conditions under which these phase changes occur is crucial. For example, water boils at 100 degrees Celsius at sea level, but this temperature changes with altitude. Similarly, the freezing point of water is 0 degrees Celsius, but it can be influenced by the presence of impurities.

In this subchapter, we will discuss the energy changes associated with phase changes, such as heat of fusion and heat of vaporization. We will also delve into the effects of pressure and temperature on the behavior of substances during phase changes.

By studying the changes in states of matter, students will gain a solid foundation in understanding the behavior of different substances. This knowledge will be invaluable in various fields, such as materials science, environmental science, and chemical engineering. Understanding these concepts will help students make connections between the macroscopic world they observe and the microscopic world of atoms and molecules.

Phase Diagrams: Understanding Phase Transitions

In the fascinating world of chemistry, understanding phase transitions is essential for comprehending the behavior of matter under different conditions. Phase transitions refer to the changes that occur when a substance transitions from one physical state to another, such as from solid to liquid or liquid to gas. To grasp these transformations, scientists use a valuable tool called phase diagrams. These diagrams provide a visual representation of the different phases a substance can exist in under specific combinations of temperature and pressure.

In this subchapter, we will delve into the realm of phase diagrams, unraveling their secrets and demonstrating how they aid in our comprehension of phase transitions. By the end, you will have a solid grasp of these diagrams and their significance in the field of chemistry.

Phase diagrams are structured with temperature on the x-axis and pressure on the y-axis. They are divided into regions that represent different phases, such as solid, liquid, and gas. The boundaries between these regions, known as phase boundaries, indicate the conditions at which phase transitions occur.

One important concept illustrated by phase diagrams is the triple point. This point represents the unique combination of pressure and temperature at which all three phases coexist in equilibrium. It serves as a reference point for understanding phase transitions.

Moreover, phase diagrams allow us to explore critical points, which define the conditions required to transform a substance from a gas to a liquid or from a liquid to a solid. At critical points, the distinction between phases blurs, and the substance exhibits unique properties.

Understanding phase diagrams is essential for predicting the behavior of substances under different conditions. By analyzing a phase diagram, we can determine the conditions required to induce a phase transition and identify the stability of a substance in different states.

In conclusion, phase diagrams are an indispensable tool for comprehending phase transitions in chemistry. By examining these diagrams, we can visualize and understand how substances transform from one phase to another based on temperature and pressure. The knowledge gained from studying phase diagrams enables us to predict and control the behavior of matter, leading to numerous scientific advancements. So, dive into this subchapter, and let's unravel the mysteries of phase diagrams together!

Chapter 6: Chemical Reactions

Introduction to Chemical Reactions

Chemistry is the study of matter and its properties, and one of the most fascinating aspects of this field is chemical reactions. In this subchapter, we will delve into the world of chemical reactions, exploring the fundamental principles and concepts that govern these transformations. Whether you are a high school student embarking on your chemistry journey or a curious individual with a passion for science, this introduction will provide you with a solid foundation to understand the magic happening at the molecular level.

Chemical reactions are the driving force behind countless natural and artificial processes that occur around us every day. From the rusting of iron to the combustion of fuel in an engine, chemical reactions are responsible for the changes we observe in the world. Understanding these reactions not only allows us to explain these phenomena but also enables us to harness their power for practical applications.

At its core, a chemical reaction involves the transformation of one or more substances, called reactants, into different substances, known as products. This transformation occurs due to the breaking and forming of chemical bonds between atoms, leading to the rearrangement of atoms and the creation of new compounds. The reactants and products in a chemical reaction are represented by chemical equations, which provide a concise way to communicate the details of the reaction.

Chemical reactions can be classified into various types, including synthesis reactions, decomposition reactions, combustion reactions, and more. Each type has its unique characteristics and underlying principles. By understanding these different types, you will gain insight into the patterns and behaviors that govern chemical transformations.

Moreover, chemical reactions are governed by fundamental principles, such as the law of conservation of mass and the law of definite proportions. These laws dictate that matter is neither created nor destroyed in a chemical reaction and that the ratio of elements in a compound is constant. Understanding these principles will help you make sense of chemical equations and predict the outcomes of reactions.

In this subchapter, we will explore these concepts in detail, providing clear explanations, illustrative examples, and practical applications. By the end, you will have a solid understanding of the basics of chemical reactions, setting the stage for further exploration into the fascinating world of chemistry.

So, let's embark on this exciting journey into the realm of chemical reactions, where atoms dance, bonds break, and new substances are formed. Get ready to unlock the secrets of the elements and witness the magic of chemistry unfold before your eyes.

Balancing Chemical Equations

Chemical equations are an essential part of understanding the reactions that occur between different substances. They provide a concise representation of the reactants and products involved in a chemical reaction. However, these equations need to be balanced to accurately depict the conservation of mass during a reaction. In this subchapter, we will delve into the art of balancing chemical equations and unravel the secrets behind this crucial skill in chemistry.

To begin, let's define what balancing a chemical equation means. Balancing involves adjusting the coefficients of the reactants and products so that the number of atoms of each element is the same on both sides of the equation. This ensures that the law of conservation of mass is obeyed – matter cannot be created or destroyed during a chemical reaction.

Balancing chemical equations requires careful attention to detail and a systematic approach. The first step is to identify the different elements present in the equation. Then, count the number of atoms of each element on both sides of the equation and compare them. If the numbers are unequal, you need to adjust the coefficients to achieve balance.

To help you balance equations effectively, there are a few strategies you can employ. Start by balancing the elements that appear in the fewest compounds first. This simplifies the process and prevents errors. Remember to avoid changing the subscripts in the formulas, as this would alter the identity of the compound.

In some cases, you may encounter polyatomic ions, which are groups of atoms with a charge. Treat these ions as single entities and balance them as a whole. Additionally, be mindful of balancing the charges in the equation to ensure overall charge neutrality.

Balancing chemical equations is not just a mathematical exercise; it is a key tool for understanding the underlying chemistry. By balancing equations, you gain insight into the stoichiometry of the reaction, such as the mole ratios between reactants and products. This information is crucial for determining the amount of reactants needed or the yield of products in a chemical reaction.

Practicing the art of balancing chemical equations is vital for success in chemistry. It enhances your problem-solving skills, logical thinking, and ability to interpret chemical phenomena. So grab your pencil and start mastering this fundamental skill that will unlock the mysteries of chemical reactions and pave the way for your understanding of the elements.

Types of Chemical Reactions

Chemistry is a fascinating subject that helps us understand the world around us. One of the fundamental aspects of chemistry is the study of chemical reactions. Chemical reactions occur when substances combine or change into new substances through the breaking and forming of chemical bonds. In this subchapter, we will explore the different types of chemical reactions that exist.

1. Combination Reactions: In combination reactions, two or more substances combine to form a single product. For example, when hydrogen gas reacts with oxygen gas, water is formed.

2. Decomposition Reactions: Decomposition reactions are the opposite of combination reactions. In these reactions, a single compound breaks down into two or more simpler substances. An example of a decomposition reaction is when hydrogen peroxide breaks down into water and oxygen gas.

3. Displacement Reactions: Displacement reactions involve the exchange of atoms or ions between two different compounds. There are two types of displacement reactions: single displacement and double displacement. In single displacement reactions, one element replaces another element in a compound. For instance, when zinc reacts with hydrochloric acid, it displaces hydrogen and forms zinc chloride. Double displacement reactions occur when the cations or anions of two compounds exchange places. An example is the reaction between sodium chloride and silver nitrate, which produces sodium nitrate and silver chloride.

4. Acid-Base Reactions: Acid-base reactions, also known as neutralization reactions, occur when an acid reacts with a base to produce a salt and water. For instance, when hydrochloric acid reacts with sodium hydroxide, sodium chloride and water are formed.

5. Redox Reactions: Redox reactions involve the transfer of electrons between reactants. Oxidation is the loss of electrons, while reduction is the gain of electrons. These reactions are crucial in the study of energy production and storage, as well as in the field of electrochemistry.

Understanding the different types of chemical reactions is essential for students to comprehend the underlying principles of chemistry. By recognizing the patterns and characteristics of each type, students can predict and interpret the outcomes of various reactions. This knowledge is not only valuable for academic purposes but also for practical applications, such as in the fields of medicine, engineering, and environmental science.

In the next subchapter, we will delve deeper into the concept of chemical equations and how they are used to represent and balance these reactions.

Stoichiometry: Calculating Reactant and Product Amounts

In the fascinating world of chemistry, stoichiometry is a key concept that allows us to understand the relationships between reactants and products in a chemical reaction. If you've ever wondered how scientists calculate the amount of reactants needed or the amount of products formed, then this subchapter is for you!

Stoichiometry is all about using balanced chemical equations to determine the relative amounts of substances involved in a reaction. It helps us answer questions like: How much product can be formed from a given amount of reactants? How much reactant is needed to produce a certain amount of product? These calculations are crucial for scientists to predict the outcome of a reaction and optimize the efficiency of chemical processes.

To get started with stoichiometry, you need to understand the concept of a balanced chemical equation. A balanced equation shows the exact number of molecules or moles of each substance involved in a reaction. It ensures that the law of conservation of mass is upheld, meaning that the total mass of the reactants equals the total mass of the products.

Using the coefficients in a balanced equation, we can convert between the number of moles of each substance involved. This allows us to calculate reactant and product amounts. For example, if we know the number of moles of one reactant, we can use the stoichiometric ratio to determine the number of moles of another reactant or the amount of product formed.

In this subchapter, we will explore various methods to solve stoichiometry problems, including dimensional analysis and the use of molar ratios. We will provide step-by-step examples and practice exercises to improve your understanding and problem-solving skills.

By mastering stoichiometry, you will gain a deeper understanding of chemical reactions and their quantitative aspects. Whether you are preparing for an exam, pursuing a career in science, or simply curious about the world around you, this subchapter will equip you with the tools to tackle stoichiometry problems with confidence.

So, join us on this exciting journey through stoichiometry, and unlock the secrets behind calculating reactant and product amounts. Get ready to make chemistry easy and enjoyable!

Chapter 7: Acids, Bases, and pH

Properties of Acids and Bases

In the world of chemistry, acids and bases are fundamental substances that play a crucial role in various chemical reactions. Understanding the properties of acids and bases is essential for any student exploring the fascinating world of chemistry. This subchapter aims to provide a comprehensive overview of the properties of acids and bases, helping students grasp their characteristics and behavior.

Acids are substances that release hydrogen ions (H+) when dissolved in water. One of the most prominent properties of acids is their ability to turn blue litmus paper red. This reaction occurs due to the high concentration of hydrogen ions present in acids. Additionally, acids have a sour taste, which can be observed in everyday life through substances like lemon juice or vinegar.

Another key property of acids is their corrosive nature. Acids have the ability to react with and destroy certain metals, as well as damage organic materials such as clothing or skin. It is important to handle acids with caution and use appropriate safety measures when working with them.

On the other hand, bases are substances that release hydroxide ions (OH-) when dissolved in water. Bases have the opposite effect of acids on litmus paper, turning it from red to blue. This property is due to the presence of hydroxide ions, which have a basic or alkaline nature.

One of the most well-known bases is sodium hydroxide, commonly known as lye. It is used in various products, including soaps and household cleaners. Bases have a bitter taste and slippery texture, which can be observed when touching substances like baking soda.

An essential property of bases is their ability to neutralize acids. This reaction is known as a neutralization reaction, where an acid and a base combine to form water and a salt. This property is often utilized in various applications, such as antacids to relieve stomach acid or in laboratories to neutralize harmful acidic spills.

Understanding the properties of acids and bases is crucial in many scientific fields. It lays the foundation for understanding chemical reactions, pH levels, and the behavior of various substances in different environments. By comprehending these properties, students can gain a deeper understanding of the elements and their interactions, enabling them to explore the world of chemistry more effectively.

In conclusion, the properties of acids and bases are vital for students delving into the realm of chemistry. Acids release hydrogen ions, turn litmus paper red, and have a sour taste, while bases release hydroxide ions, turn litmus paper blue, and have a bitter taste. Both acids and bases possess distinct characteristics that allow them to interact with other substances and play significant roles in chemical reactions. By understanding these properties, students can unlock the secrets of the elements and pave the way for a deeper understanding of the world of chemistry.

pH Scale: Measuring Acidity and Basicity

Understanding the concept of acidity and basicity is essential to grasp the fundamentals of chemistry. The pH scale is a powerful tool that allows us to measure and quantify the acidity or basicity of a substance. In this subchapter, we will explore the pH scale and learn how it helps us understand the properties of various substances.

The pH scale is a numerical scale that ranges from 0 to 14, with 7 being considered neutral. Substances with a pH below 7 are acidic, while those with a pH above 7 are basic. This scale provides a standardized way to compare the relative acidity or basicity of different solutions.

To determine the pH of a substance, we use a device called a pH meter or pH indicator paper. These tools provide a quick and accurate measurement of a solution's pH level. The pH meter uses a glass electrode that generates a small electrical voltage when it comes into contact with the solution. This voltage is then converted into a pH value by the device.

Acidity and basicity are determined by the concentration of hydrogen ions (H+) and hydroxide ions (OH-) in a solution. In an acidic solution, there is a higher concentration of H+ ions, while in a basic solution, there is a higher concentration of OH- ions. As the concentration of H+ ions increases, the pH value decreases, indicating a stronger acid. Conversely, as the concentration of OH- ions increases, the pH value increases, indicating a stronger base.

Understanding the pH scale is crucial in many scientific fields, including environmental science, medicine, and biochemistry. For instance, in environmental science, the pH of water bodies is crucial to determine their suitability for aquatic life. In medicine, the pH of blood is closely regulated by the body to ensure proper physiological functioning.

In conclusion, the pH scale is a vital tool that enables us to measure and understand the acidity and basicity of various substances. By using pH meters or indicators, we can quantitatively determine the pH value of a solution. This knowledge is invaluable in many scientific disciplines and provides a foundation for further exploration of chemistry and its applications. As students, mastering the pH scale will equip us with a powerful tool to decipher the intricacies of the chemical world.

Acid-Base Reactions and Neutralization

In the fascinating world of chemistry, there are various types of reactions that occur between different substances. One such type is acid-base reactions, which play a crucial role in our everyday lives. Understanding these reactions and the process of neutralization is essential for students studying chemistry.

Acid-base reactions involve the interaction between an acid and a base, resulting in the formation of a salt and water. Acids are substances that can donate protons (H+) when dissolved in water, while bases are substances that can accept protons. The reaction between an acid and a base is often referred to as neutralization because it results in the formation of a neutral product.

The key concept behind acid-base reactions is the concept of pH. The pH scale measures the acidity or alkalinity of a solution. Acids have a pH value less than 7, while bases have a pH value greater than 7. A neutral solution, such as pure water, has a pH of 7.

During a neutralization reaction, the acid and base combine to form water and a salt. The salt is formed when the positive ion from the base combines with the negative ion from the acid. For example, when hydrochloric acid (HCl) reacts with sodium hydroxide (NaOH), the resulting products are water (H2O) and sodium chloride (NaCl).

Neutralization reactions have numerous applications in our daily lives. One such example is the use of antacids to relieve heartburn. Antacids contain basic substances that neutralize excess stomach acid, providing relief to individuals suffering from indigestion.

Understanding acid-base reactions and neutralization is crucial not only in chemistry but also in fields like medicine and environmental science. It helps us understand the behavior of various substances and their effects on our bodies and the environment.

To summarize, acid-base reactions and neutralization play a vital role in the world of chemistry. They involve the interaction between an acid and a base, resulting in the formation of a salt and water. Understanding these reactions and their applications is essential for students studying chemistry, as it provides a foundation for further exploration in various scientific fields. So, dive into the world of acid-base reactions and discover the wonders of chemistry!

Chapter 8: Organic Chemistry

Introduction to Organic Chemistry

Organic chemistry is a fascinating branch of chemistry that focuses on the study of carbon-based compounds. It is a vital field that plays a crucial role in various aspects of our daily lives, from the medicines we take to the materials we use. In this subchapter, we will delve into the fundamentals of organic chemistry, providing you, the student, with a solid understanding of the key concepts and principles.

To begin our journey, let's explore the unique characteristics of organic compounds. Unlike inorganic compounds, which primarily consist of metals and nonmetals, organic compounds contain carbon atoms bonded with other elements such as hydrogen, oxygen, nitrogen, and more. This ability of carbon to form numerous bonds makes it the backbone of organic chemistry.

Understanding the structure of organic compounds is essential in deciphering their properties and reactions. We will introduce you to the concept of functional groups, which are specific arrangements of atoms that determine the chemical behavior of a compound. From alcohols to carboxylic acids, you will learn to recognize and classify different functional groups.

Next, we will delve into the nomenclature of organic compounds. You will discover the systematic rules for naming organic molecules, which allows scientists to communicate unambiguously. We will guide you through the process of naming alkanes, alkenes, and alkynes, providing you with practical examples and exercises to reinforce your understanding.

As we progress, the subchapter will introduce you to the fundamental reactions and mechanisms in organic chemistry. You will explore topics such as substitution, addition, elimination, and oxidation-reduction reactions. By comprehending these reactions and their underlying mechanisms, you will gain the ability to predict and understand the behavior of organic compounds.

Moreover, we will discuss the significance of organic chemistry in real-world applications. From the synthesis of pharmaceutical drugs to the development of sustainable materials, you will discover how organic chemistry impacts various scientific and technological advancements. This knowledge will not only enhance your understanding of the subject but also highlight its relevance in addressing global challenges.

In conclusion, this subchapter serves as an introduction to the captivating world of organic chemistry. By providing a solid foundation in the key concepts, structures, reactions, and applications, we aim to equip you, the student, with the necessary tools to navigate through the complexities of this discipline. Whether you aspire to pursue a career in science or simply wish to expand your knowledge, understanding organic chemistry is essential for unraveling the wonders of the elements that surround us.

Hydrocarbons: Alkanes, Alkenes, and Alkynes

Welcome to the exciting world of hydrocarbons! In this subchapter, we will explore the fascinating properties and characteristics of three important types of hydrocarbons: alkanes, alkenes, and alkynes. Understanding these compounds is crucial for any budding chemist, as they form the building blocks of organic chemistry and have numerous practical applications.

Let's start with alkanes, which are the simplest type of hydrocarbon. They consist of only single bonds between carbon atoms and are often referred to as "saturated hydrocarbons." Alkanes have a general formula of C_nH_{2n+2}, where n represents the number of carbon atoms. Methane, ethane, and propane are some examples of alkanes commonly found in nature. They are often used as fuels due to their high energy content.

Moving on to alkenes, we encounter a more complex structure. Alkenes contain at least one double bond between carbon atoms, which gives them a distinctive reactivity. The general formula for alkenes is C_nH_{2n}, and they are commonly known as "unsaturated hydrocarbons." Ethene and propene are examples of alkenes that have numerous industrial applications, such as in the production of plastics and synthetic fibers.

Finally, we come to alkynes, the most reactive type of hydrocarbon. Alkynes contain at least one triple bond between carbon atoms and have the general formula C_nH_{2n-2}. They are often called "acetylenes" and are used in various industrial processes, including the production of rubber and plastics. Ethyne, commonly known as acetylene, is the simplest alkyne and is widely used in welding and cutting torches.

Understanding the properties and behaviors of alkanes, alkenes, and alkynes is crucial for predicting their reactivity and behavior in different chemical reactions. Furthermore, their structural differences directly impact their physical properties, such as boiling points and solubilities. By studying these hydrocarbons, you will gain a deeper appreciation for the complexity and diversity of organic compounds.

In conclusion, hydrocarbons play a fundamental role in chemistry and have significant practical applications. Alkanes, alkenes, and alkynes represent different classes of hydrocarbons with distinct structural features and reactivities. By mastering the concepts discussed in this subchapter, you will be well-equipped to tackle more advanced topics in organic chemistry and explore the limitless possibilities of this fascinating field.

Functional Groups: Alcohols, Aldehydes, and Ketones

In the exciting world of chemistry, functional groups play a vital role in determining the properties and behaviors of organic compounds. In this subchapter, we will delve into the fascinating world of three important functional groups: alcohols, aldehydes, and ketones.

First up, let's explore alcohols. These compounds contain a hydroxyl (-OH) group bonded to a carbon atom. Alcohols are commonly found in everyday life, from the ethanol in alcoholic beverages to the isopropyl alcohol used as a disinfectant. They can be classified as either primary, secondary, or tertiary, depending on the number of carbon atoms bonded to the carbon bearing the hydroxyl group. Alcohols display unique chemical properties, such as the ability to undergo oxidation reactions, forming aldehydes or ketones.

Next, we move on to aldehydes, which contain a carbonyl group (C=O) at the end of a carbon chain. Aldehydes are known for their distinctive odor and are often used as flavoring agents or fragrances. Formaldehyde, a common aldehyde, is widely used in the production of plastics and resins. Aldehydes can be further oxidized to carboxylic acids, making them essential intermediates in various organic reactions.

Lastly, we encounter ketones, which also possess a carbonyl group (C=O) but are located within a carbon chain. Ketones are commonly used as solvents and are found in many household products, including nail polish remover and paint thinners. Their unique structure gives them different chemical properties than aldehydes, making them valuable in a range of applications.

Understanding the properties and reactions of alcohols, aldehydes, and ketones is crucial in many scientific fields, including medicine, pharmaceuticals, and biochemistry. It allows scientists to design and synthesize new compounds with specific functions, contributing to advancements in drug development and materials science.

As students, it is essential to grasp the fundamentals of these functional groups and their respective properties. By doing so, we can better comprehend the structure-function relationship of organic compounds, and gain a deeper understanding of the complex world of chemistry.

In conclusion, alcohols, aldehydes, and ketones are three significant functional groups in organic chemistry. Their unique structures and properties make them indispensable in various industries and scientific research. By studying these functional groups, we can unravel the mysteries of organic compounds and unlock a world of possibilities in the field of science education.

Organic Reactions: Substitution, Addition, and Elimination

Welcome to the fascinating world of organic reactions! In this subchapter, we will explore three fundamental types of reactions: substitution, addition, and elimination. Understanding these reactions is crucial to mastering the principles of organic chemistry and unraveling the mysteries of the elements. So let's dive in!

Substitution reactions are the cornerstone of organic chemistry. They occur when one atom or group is replaced by another in a molecule. Think of it as a game of molecular musical chairs, with atoms constantly swapping places. This reaction type is commonly observed in alkyl halides, where a halogen atom is replaced by another atom or group. Substitution reactions are classified into two main categories: nucleophilic and electrophilic. Nucleophilic substitution involves the attack of a nucleophile on an electrophilic carbon, leading to the displacement of the leaving group. On the other hand, electrophilic substitution occurs when an electrophile reacts with an aromatic compound, resulting in the substitution of a hydrogen atom.

Moving on to addition reactions, these occur when two molecules combine to form a larger molecule. It's like adding pieces to a puzzle! Addition reactions are commonly observed in alkenes and alkynes, where the carbon-carbon double or triple bond is broken and replaced with new atoms or groups. One classic example is the addition of hydrogen gas to an alkene, resulting in the formation of an alkane. Another important addition reaction is the hydration of an alkene, where water adds across the double bond to form an alcohol.

Last but not least, we have elimination reactions. These reactions involve the removal of atoms or groups from a molecule, resulting in the formation of a double or triple bond. It's like a molecular Houdini act! Elimination reactions are commonly observed in alkyl halides and alcohols. One well-known example is the dehydration of an alcohol, where water is eliminated to form an alkene. Another important elimination reaction is the dehydrohalogenation of an alkyl halide, where a hydrogen halide is removed to form an alkene.

Understanding organic reactions is essential for unraveling the complexity of the elements. By grasping the concepts of substitution, addition, and elimination, you will be able to predict and explain the behavior of organic compounds. So buckle up, students, and get ready to embark on a thrilling journey through the world of organic reactions. Let's unlock the secrets of the elements together!

Chapter 9: Chemical Kinetics

Introduction to Chemical Kinetics

Chemical kinetics is an essential branch of chemistry that deals with the study of reaction rates and the factors that influence them. In this subchapter, we will explore the fascinating world of chemical kinetics and understand how reactions occur and progress at the molecular level.

Chemical kinetics provides us with valuable insights into the speed and mechanism of chemical reactions. By investigating the rate at which reactants are consumed and products are formed, we can gain a deeper understanding of the factors that affect the overall reaction rate. This knowledge is crucial in various scientific fields, including pharmaceuticals, environmental sciences, and material science.

Understanding chemical kinetics requires a basic grasp of reaction rates, which measure the change in concentration of reactants or products per unit time. We will delve into the concept of rate laws and explore how they are determined experimentally. Additionally, we will learn about rate constants and how they relate to reaction rates.

Factors such as temperature, concentration, and catalysts play significant roles in influencing reaction rates. We will discuss their effects on chemical reactions and examine the principles behind them. Furthermore, we will explore collision theory, which explains how reactant molecules must collide with sufficient energy and proper orientation to form products.

Chemical kinetics also involves the study of reaction mechanisms, which are step-by-step processes that describe how reactants transform into products. By understanding reaction mechanisms, we can predict the intermediates formed during a reaction and identify the rate-determining step, which controls the overall reaction rate.

In this subchapter, we will learn about various mathematical models, such as integrated rate laws, which allow us to determine the concentration of reactants or products at any given time during a reaction. This knowledge is essential for predicting reaction outcomes and designing chemical processes.

Chemical kinetics is a dynamic field that continually contributes to scientific advancements. By mastering the concepts and principles discussed in this subchapter, you will be equipped with the fundamental knowledge necessary to explore the intricacies of chemical reactions and their rates. Join us on this exciting journey through the world of chemical kinetics and unlock the secrets behind how reactions occur and progress.

Whether you are a student pursuing a career in science or simply have a passion for understanding the elements, this subchapter will provide you with a solid foundation in chemical kinetics. Get ready to embark on a captivating exploration of reaction rates and their underlying principles. Let's dive in!

Reaction Rates: Factors Affecting Speed

Understanding the factors that affect the speed of reactions is crucial in the study of chemistry. In this subchapter, we will explore the various factors that influence reaction rates and how they can be manipulated to control chemical processes. Whether you're a chemistry student or simply interested in science education, this section will provide you with valuable insights into the fascinating world of reactions.

First and foremost, it's important to grasp the concept of reaction rates. Reaction rate refers to how fast a reaction occurs, specifically the speed at which reactants are converted into products. Several factors can influence this speed, and one of the most significant is the concentration of reactants. Generally, an increase in reactant concentration leads to a higher reaction rate, as more particles are available to collide and form products. We will delve into the mathematical relationship between concentration and reaction rates, as well as the concept of rate laws.

Another crucial factor affecting reaction rates is temperature. Increasing the temperature usually accelerates the rate of reaction, as higher temperatures provide more energy to the reacting particles. We will explore the relationship between temperature and reaction rates, including the activation energy required for a reaction to occur.

Furthermore, the presence of catalysts can significantly influence reaction rates. Catalysts are substances that speed up reactions by providing an alternative reaction pathway with a lower activation energy. We will investigate the mechanism by which catalysts function and their importance in various industrial and biological processes.

Surface area is another factor that affects reaction rates. Increasing the surface area of a solid reactant exposes more particles to the other reactants, leading to more frequent collisions and faster reactions. We will discuss the significance of surface area and its practical applications.

Lastly, we will explore the impact of pressure on reaction rates, particularly in gaseous reactions. Increasing pressure forces gas molecules to occupy a smaller space, resulting in more frequent collisions and faster reaction rates.

Understanding the factors that affect reaction rates is crucial for predicting and controlling chemical reactions. By manipulating these factors, scientists and engineers can optimize chemical processes in various fields, ranging from pharmaceuticals to environmental studies. Whether you're a student or a science enthusiast, this subchapter will equip you with the knowledge needed to comprehend the intricacies of reaction rates and their practical implications.

Chemical Equilibrium: Balancing Forward and Reverse Reactions

In the fascinating world of chemistry, one of the most important concepts to grasp is chemical equilibrium. It is the state at which the forward and reverse reactions in a chemical system occur at the same rate, resulting in a dynamic balance. Understanding this concept is crucial for students who wish to delve deeper into the complexities of chemical reactions.

Chemical equilibrium occurs when the concentrations of reactants and products remain constant over time. This state is achieved when the forward reaction, which transforms reactants into products, is counterbalanced by the reverse reaction, where products react to form reactants. Although it might seem like these reactions have come to a halt, they are actually happening continuously, leading to a stable equilibrium.

To understand chemical equilibrium fully, it is essential to comprehend the concept of the equilibrium constant (K). This constant is a numerical value that represents the relative concentrations of reactants and products at equilibrium. By knowing the equilibrium constant, we can predict the direction in which a reaction will proceed or determine the concentrations of reactants and products at equilibrium using the initial concentrations.

Le Chatelier's principle is another crucial concept related to chemical equilibrium. It states that when a system in equilibrium is subjected to a change in temperature, pressure, or concentration, it will respond by shifting the equilibrium to counteract the change. For example, if the concentration of a reactant is increased, the equilibrium will shift towards the product side to restore balance.

Chemical equilibrium plays a vital role in various fields of science, including environmental chemistry, biochemistry, and chemical engineering. It is essential for understanding complex systems such as the behavior of gases, the formation of acids and bases, and the dynamics of enzyme-catalyzed reactions.

By grasping the concept of chemical equilibrium, students can gain a deeper understanding of the factors that influence the progress of a reaction and the conditions required for equilibrium. This knowledge opens doors to exploring advanced topics such as acid-base equilibria, solubility equilibria, and the intricacies of chemical kinetics.

In conclusion, chemical equilibrium is a fundamental concept in chemistry that allows us to understand the dynamic balance between forward and reverse reactions. By studying this concept, students can gain a deeper understanding of the factors influencing chemical systems and their behavior. Chemical equilibrium is not only essential for academic success but also plays a crucial role in various scientific fields. So, let's dive into the fascinating world of chemical equilibrium and unlock the mysteries of balancing forward and reverse reactions!

Chapter 10: Electrochemistry

Introduction to Electrochemistry

Electrochemistry is a fascinating branch of chemistry that focuses on the study of the relationship between electricity and chemical reactions. It explores how electric current can be used to drive or facilitate chemical reactions, and how chemical reactions can generate electrical energy. This subchapter will introduce you to the fundamental concepts and principles of electrochemistry, providing a solid foundation for further exploration in this exciting field.

Electrochemical reactions occur through the transfer of electrons between species, which are either atoms, ions, or molecules. These reactions take place in devices called electrochemical cells, where the reactants are contained in separate compartments connected by a conductive pathway. The two main types of electrochemical cells are galvanic (voltaic) cells and electrolytic cells.

Galvanic cells, also known as batteries, are devices that convert chemical energy into electrical energy. They consist of two half-cells, each containing an electrode immersed in an electrolyte solution. During the reaction, electrons flow from the anode (the electrode where oxidation occurs) to the cathode (the electrode where reduction occurs) through an external circuit, providing a source of electrical energy.

On the other hand, electrolytic cells are used to drive non-spontaneous chemical reactions through the application of an external electric current. These cells require an input of electrical energy to force the reaction to occur. Electrolysis is a common application of electrolytic cells and is used for various purposes, such as electroplating, metal extraction, and water splitting.

Understanding the principles of electrochemistry is crucial for many real-world applications. It plays a vital role in the development of batteries for portable devices, electric vehicles, and renewable energy storage. Moreover, electrochemical techniques are widely used in analytical chemistry for the detection and quantification of chemical species, making it a valuable tool in environmental monitoring and medical diagnostics.

In this subchapter, we will discuss key concepts such as oxidation-reduction reactions, half-cell reactions, cell potentials, and the Nernst equation. We will also explore the different types of electrochemical cells and their applications. By the end of this subchapter, you will have a solid understanding of the fundamental principles that underpin electrochemistry, enabling you to delve deeper into this fascinating field.

Whether you are a chemistry enthusiast or a student pursuing a career in science, this subchapter will provide you with a comprehensive introduction to electrochemistry. Get ready to unravel the mysteries of electricity and chemical reactions as we embark on this electrifying journey through the world of electrochemistry!

Redox Reactions: Oxidation and Reduction

In the fascinating world of chemistry, one of the most fundamental concepts is the phenomenon of redox reactions. These reactions, also known as oxidation-reduction reactions, play a crucial role in our daily lives, from the rusting of iron to the process of respiration in our bodies. Understanding the principles behind redox reactions is essential for any student of chemistry, as it forms the basis for numerous other concepts in this captivating field.

So, what exactly are redox reactions? Well, they involve the transfer of electrons between two species, leading to changes in their oxidation states. To grasp this concept fully, it is important to understand the terms oxidation and reduction. Oxidation refers to the loss of electrons by a species, resulting in an increase in its oxidation state. On the other hand, reduction involves the gain of electrons by a species, leading to a decrease in its oxidation state.

To simplify the understanding of redox reactions, it is common to use the mnemonic "LEO the lion says GER" (Loss of Electrons is Oxidation, Gain of Electrons is Reduction). This catchy phrase helps students remember the key concepts of redox reactions.

One classic example of a redox reaction is the rusting of iron. When iron comes into contact with oxygen and moisture, it undergoes oxidation, resulting in the formation of iron(III) oxide, commonly known as rust. This process is not only a nuisance when it comes to the appearance of our beloved metal objects but also a reminder of the fundamental principles of redox reactions.

Redox reactions are not limited to chemical processes alone; they also play a vital role in biological systems. For instance, during respiration, glucose is oxidized to produce energy while oxygen is reduced to form water. This intricate process is essential for our survival and highlights the importance of redox reactions in the realm of science and biology.

In conclusion, redox reactions are a fundamental aspect of chemistry that students must grasp to understand the elements fully. These reactions involve the transfer of electrons between species, resulting in changes in oxidation states. Remembering the mnemonic "LEO the lion says GER" can help students differentiate between oxidation and reduction. From the rusting of iron to the process of respiration, redox reactions are pervasive in our daily lives and have a significant impact on both the chemical and biological realms. So, dive into the world of redox reactions and unlock the key to understanding the elements!

Electrolysis: Breaking Down Compounds with Electricity

Electrolysis is a fascinating process that allows us to break down compounds into their individual elements using the power of electricity. In this subchapter, we will explore the concept of electrolysis and how it revolutionized the field of chemistry.

To begin with, let's understand the basic principles behind electrolysis. At its core, electrolysis involves the use of an electrical current to drive a non-spontaneous chemical reaction. This means that we can break apart compounds into their constituent elements by passing electricity through them.

One of the most notable applications of electrolysis is the extraction of metals from their ores. For instance, the extraction of aluminum from bauxite ore is made possible through the process of electrolysis. By passing an electric current through a molten mixture of aluminum oxide and cryolite, we can separate the aluminum ions from the oxygen ions, resulting in the production of pure aluminum.

Electrolysis also plays a crucial role in the field of electroplating. This process involves depositing a layer of metal onto a surface to enhance its appearance or provide protection. By using electrolysis, we can coat objects with a thin layer of metal, such as gold or silver, by immersing them in a solution containing the desired metal ions and passing an electric current through it. This allows the metal ions to migrate to the object's surface and form a uniform layer.

Furthermore, electrolysis has important implications in the field of energy storage. Through a process called electrolysis of water, we can split water molecules into hydrogen and oxygen gases. This hydrogen gas can then be utilized as a clean and renewable energy source. By passing an electric current through water, hydrogen ions migrate to the cathode (negative electrode), while oxygen ions migrate to the anode (positive electrode). The resulting hydrogen gas can be stored and used as a fuel to generate electricity or power vehicles.

In conclusion, electrolysis is a powerful tool that enables us to break down compounds into their elemental components using electricity. From the extraction of metals to electroplating and energy storage, electrolysis has revolutionized various aspects of chemistry. Understanding this process opens up a world of possibilities for scientists and engineers, paving the way for advancements in science and technology.

Chapter 11: Nuclear Chemistry

Nuclear Reactions: Radioactivity and Decay

In the fascinating world of chemistry, one of the most intriguing and powerful phenomena is nuclear reactions. These reactions involve the nucleus of an atom, the core that holds the protons and neutrons. Nuclear reactions can bring about incredible changes, releasing enormous amounts of energy and transforming one element into another. This subchapter will delve into the captivating concepts of radioactivity and decay, shedding light on these fundamental processes.

Radioactivity is a natural property of certain elements, known as radioactive elements. These elements have unstable nuclei that can spontaneously undergo radioactive decay. During this process, the nucleus emits radiation in the form of alpha particles, beta particles, or gamma rays. This radiation is highly energetic and can penetrate various materials, making it crucial to understand its effects.

Alpha particles are made up of two protons and two neutrons, and they have a positive charge. Beta particles, on the other hand, can be either electrons (beta minus decay) or positrons (beta plus decay). Gamma rays are high-energy photons. The emission of these particles and rays leads to the transformation of the parent element into a different element, often with a different number of protons and neutrons.

Radioactive decay follows specific patterns, allowing scientists to predict the rate at which it occurs. The half-life of a radioactive substance is the time it takes for half of the atoms in a sample to decay. This concept is vital in understanding the stability and radioactivity of various elements and isotopes.

Beyond its scientific significance, nuclear reactions have important practical applications. For instance, radioisotopes are used in medicine for diagnostic purposes and cancer treatments. They can also be employed in industrial settings to detect leaks or study the behavior of materials under extreme conditions.

Understanding the principles of nuclear reactions is crucial in various scientific disciplines, including chemistry, physics, and environmental science. By comprehending radioactivity and decay, students gain a deeper understanding of the fundamental building blocks of matter and the forces that govern our universe. Moreover, this knowledge equips them with the tools to explore cutting-edge scientific advancements and contribute to the betterment of society.

In this subchapter, we will explore the different types of radiation, learn how to calculate half-lives, and delve into the practical applications of nuclear reactions. By the end, you will have a solid foundation in understanding the exciting world of radioactivity and decay, paving the way for further exploration in the realm of chemistry and scientific discovery.

Half-Life: Calculating Radioactive Decay

Understanding the concept of radioactive decay is crucial in the field of chemistry. It allows us to determine the rate at which a radioactive substance breaks down over time. One of the most important aspects of radioactive decay is the concept of half-life. In this subchapter, we will explore the concept of half-life and how to calculate it.

Radioactive decay occurs when the nucleus of an unstable atom undergoes a spontaneous change, releasing radiation in the process. These unstable atoms are referred to as radioactive isotopes. Each radioactive isotope has a specific half-life, which is the time it takes for half of the atoms in a sample to decay.

To calculate the half-life of a radioactive substance, we need to know the initial amount of the substance and the rate at which it decays. The rate of decay is often expressed as the decay constant, symbolized by λ. The decay constant is unique to each radioactive isotope and can be found in reference tables.

The formula to calculate the half-life of a radioactive substance is as follows:

$$t_{1/2} = (\ln 2) / \lambda$$

In this formula, $t_{1/2}$ represents the half-life, $\ln 2$ is the natural logarithm of 2 (approximately 0.693), and λ is the decay constant. By plugging in the values for λ, we can determine the half-life of a specific radioactive isotope.

It is important to note that the half-life remains constant regardless of the initial amount of the substance. This characteristic makes half-life particularly useful in various scientific fields, such as archaeology, geology, and medicine.

Understanding half-life calculations also allows us to determine the amount of a radioactive substance remaining after a certain period. By using the equation:

$$N = N_0 * (1/2)^{(t / t_{1/2})}$$

where N represents the remaining amount of the substance, N_0 is the initial amount, t is the time passed, and $t_{1/2}$ is the half-life, we can calculate the amount of a radioactive substance at any given time.

Calculating half-life is a fundamental skill in chemistry and provides valuable insights into the behavior of radioactive isotopes. By mastering this concept, students can better understand the principles of radioactivity and its applications in various scientific disciplines.

In conclusion, the concept of half-life is a vital aspect of understanding radioactive decay. By calculating half-life, we can determine the rate at which a substance decays and how much remains over time. This knowledge is essential for students studying chemistry and provides a solid foundation for further exploration in the field of science education.

Applications of Nuclear Chemistry

Nuclear chemistry is a fascinating branch of science that deals with the study of atomic nuclei and the changes they undergo. While it may seem complex, this subchapter aims to simplify the applications of nuclear chemistry and highlight its significance in various fields. Whether you are a student of chemistry or simply curious about the applications of this intriguing subject, read on to discover its real-world applications.

1.   Medicine:
One of the most crucial applications of nuclear chemistry is in the field of medicine. Nuclear medicine utilizes radioactive isotopes to diagnose and treat various diseases. Techniques like positron emission tomography (PET) and single-photon emission computed tomography (SPECT) help doctors visualize and map the body's functioning, aiding in the diagnosis of conditions such as cancer, heart diseases, and neurological disorders. Additionally, radioisotopes are used in radiation therapy to target and destroy cancer cells while minimizing damage to healthy cells.

2.   Energy   Production:
Nuclear chemistry plays a pivotal role in generating electricity through nuclear power plants. These plants use a process called nuclear fission, where the nucleus of an atom is split, releasing a tremendous amount of energy. This energy is harnessed to generate electricity on a large scale. Nuclear power is a reliable, eco-friendly alternative to traditional fossil fuel-based energy sources, as it produces no greenhouse gases and has a significantly higher energy density.

3.   Archaeology   and   Geology: Radioactive dating methods, such as carbon-14 dating, are employed in archaeology and geology to determine the age of artifacts and rocks, respectively. By measuring the decay of radioactive isotopes present in these materials, scientists can estimate their age with remarkable accuracy. This technique has been instrumental in understanding human history, ancient civilizations, and Earth's geological processes.

4.   Food   and   Agriculture: Nuclear chemistry also finds applications in the food and agriculture industry. By exposing seeds and crops to controlled amounts of radiation, scientists can induce desirable genetic mutations, leading to the development of new plant varieties with enhanced characteristics. This process, known as mutation breeding, has been used to create disease-resistant crops, improve yield, and enhance the nutritional content of food.

These are just a few examples of the wide-ranging applications of nuclear chemistry. From medicine to energy production and from archaeology to agriculture, this branch of science has revolutionized various fields. By understanding the principles and applications of nuclear chemistry, students can appreciate its significance in solving real-world problems and contribute to the advancement of science.

In conclusion, nuclear chemistry has proven to be an indispensable tool in various disciplines. Its applications in medicine, energy production, archaeology, and agriculture have revolutionized the way we diagnose diseases, generate electricity, understand our history, and improve food production. By delving into the world of nuclear chemistry, students can unlock a deeper understanding of the elements and their role in shaping our world.

Chapter 12: The Role of Chemistry in Everyday Life

Chemistry in Medicine and Pharmaceuticals

Chemistry plays a vital role in the field of medicine and pharmaceuticals, where it is used to develop and improve drugs, understand the human body's response to medication, and ensure the safety and effectiveness of treatments. In this subchapter, we will explore the fascinating intersection of chemistry and healthcare, shedding light on how chemical principles contribute to breakthroughs in medicine.

One of the key areas where chemistry is extensively utilized is drug discovery and development. Scientists employ various chemical techniques to identify and synthesize compounds that can potentially treat diseases. Understanding the chemical properties of different compounds allows researchers to design drugs that are both effective and safe for human consumption. This involves studying the structure-activity relationship, which investigates how the structure of a drug molecule influences its efficacy and interactions with the body.

Chemistry also plays a critical role in pharmacokinetics, the study of how drugs are absorbed, distributed, metabolized, and eliminated by the body. This knowledge helps scientists determine the appropriate dosage and frequency of drug administration, ensuring that patients receive optimal treatment. Additionally, chemistry is essential in drug formulation, as it helps create different delivery methods, such as tablets, capsules, or injections, to maximize drug absorption and stability.

Furthermore, chemistry is indispensable in the field of pharmacology, which focuses on understanding how drugs interact with the body's systems and processes. This understanding enables scientists to design drugs that target specific molecules or receptors, blocking or enhancing their activity to treat various diseases. Molecular modeling and computational chemistry techniques allow researchers to simulate drug-target interactions, helping to predict and optimize drug efficacy.

Chemistry also plays a crucial role in ensuring the safety of medications. Pharmaceutical chemists analyze the chemical composition of drugs to identify impurities or potential contaminants. They also develop analytical methods to quantify the amount of active ingredient in a drug and ensure its uniformity and quality. Chemistry is also used in the study of drug metabolism and toxicology to assess the potential side effects and interactions with other substances.

In summary, the field of medicine and pharmaceuticals heavily relies on chemistry for drug discovery, development, formulation, and safety assessment. Understanding the chemical principles behind medication allows scientists to develop effective treatments and improve patient outcomes. By integrating chemistry with medical sciences, we can continue to drive advancements in healthcare and contribute to a healthier world.

Chemistry in Agriculture and Food Production

In the world of agriculture and food production, chemistry plays a significant role in ensuring the quality, safety, and sustainability of our food supply. From the fertilizers used to enhance plant growth to the preservation techniques that keep our food fresh, understanding the chemistry behind these processes is crucial.

One important aspect of chemistry in agriculture is the development and use of fertilizers. Fertilizers are essential in providing plants with the necessary nutrients to grow and thrive. Chemists analyze soil samples to determine which nutrients are lacking and then create fertilizers that contain the appropriate elements, such as nitrogen, phosphorus, and potassium, to supplement the soil. These fertilizers can be applied directly to the soil or sprayed onto plant leaves, ensuring that crops receive the nutrients they need for optimal growth.

Another area where chemistry plays a vital role is in crop protection. Pesticides and herbicides are chemical substances used to control pests and weeds that can damage crops. Chemists work to develop these compounds, ensuring that they are effective in targeting specific pests while minimizing harm to the environment and human health. Additionally, understanding the chemical properties of these substances helps scientists determine appropriate application rates and methods to maximize their efficacy.

Chemistry also plays a crucial role in food preservation and safety. Food processing techniques, such as canning and freezing, rely on the principles of chemistry to extend the shelf life of food products. Understanding chemical reactions and microbial growth allows scientists to develop preservation methods that inhibit spoilage and maintain the nutritional value of food.

Furthermore, chemistry is involved in food safety and quality control. Through chemical analysis, scientists can detect the presence of harmful substances, such as pesticides, heavy metals, and food additives, in food products. This analysis ensures that food meets regulatory standards and is safe for consumption.

As students interested in science education, understanding the chemistry behind agriculture and food production is essential. It allows us to appreciate the complex processes that occur to ensure a safe and abundant food supply. By studying the chemistry of fertilizers, pesticides, preservation techniques, and food analysis, we can contribute to the development of sustainable agricultural practices and make informed choices about the food we consume.

In conclusion, chemistry plays a critical role in agriculture and food production. It enables us to enhance crop growth, protect crops from pests and weeds, preserve food, and ensure its safety and quality. By delving into the chemistry behind these processes, students can gain a deeper understanding of the intricate relationship between chemistry, agriculture, and food production, leading to future advancements in sustainable farming practices and a better understanding of the food we consume.

Chemistry in Environmental Conservation

Chemistry plays a crucial role in environmental conservation. As students learning about the elements and their properties, it is essential to understand how chemistry can be applied to protect and preserve our environment. In this subchapter, we will explore various ways in which chemistry contributes to environmental conservation.

One of the key areas where chemistry is vital in environmental conservation is pollution control. Chemical reactions are used to neutralize harmful substances, such as acids or heavy metals, that are released into the environment. For example, in industrial processes, chemical reactions are employed to convert toxic gases into less harmful forms before releasing them into the atmosphere. This helps prevent air pollution and its detrimental effects on human health and ecosystems.

Chemistry also plays a significant role in water treatment and purification. Through chemical processes like coagulation, flocculation, and disinfection, contaminants and pollutants present in water sources can be removed or reduced. These processes involve the use of various chemicals to ensure that the water we consume is safe and free from harmful substances.

Another important aspect of environmental conservation where chemistry is utilized is waste management. Chemical reactions are employed to convert or break down hazardous waste materials into less harmful forms. Treatment processes like incineration, oxidation, and bioremediation rely on chemical reactions to transform or eliminate toxic substances, reducing their impact on the environment.

Furthermore, chemistry contributes to the development of renewable energy sources. Through the study of chemical reactions, scientists have been able to develop more efficient and sustainable energy technologies. For instance, the development of fuel cells, solar cells, and biofuels relies heavily on understanding chemical processes and reactions.

Understanding the chemistry behind climate change is also crucial for environmental conservation. Chemistry helps us comprehend the greenhouse effect, the role of greenhouse gases, and the impact of human activities on the planet's climate. By gaining a deeper understanding of these chemical processes, students can contribute to finding innovative solutions to mitigate the effects of climate change.

In conclusion, chemistry plays a vital role in environmental conservation. From pollution control and water treatment to waste management and renewable energy development, understanding the chemical processes involved helps us protect and preserve our environment. As students interested in science education, delving into the chemistry of environmental conservation not only enhances our understanding of the elements but also equips us with the knowledge to make a positive impact on our planet's future.

Chapter 13: Laboratory Techniques and Safety

Introduction to Laboratory Techniques

In the world of science education, laboratory techniques play a crucial role in the understanding of the elements. The ability to conduct experiments and analyze data in a controlled laboratory setting is essential for any aspiring scientist or chemistry student. This subchapter aims to provide an in-depth introduction to laboratory techniques, equipping students with the necessary knowledge and skills to excel in their scientific pursuits.

The first section of this subchapter explores the importance of laboratory safety. Students will learn about the potential hazards present in the laboratory and the precautions they must take to ensure their own safety and the safety of others. Topics covered include proper handling of chemicals, the use of personal protective equipment, and emergency procedures. By emphasizing the importance of safety, students will be prepared to conduct experiments responsibly and confidently.

The next section delves into the fundamental laboratory techniques that every chemistry student should be familiar with. This includes measuring volumes using various glassware, such as graduated cylinders and pipettes, and weighing substances accurately using balances. Students will also learn how to use laboratory equipment like Bunsen burners, hot plates, and magnetic stirrers. Through detailed explanations, step-by-step instructions, and visual aids, students will gain the necessary skills to perform these techniques with precision and accuracy.

Furthermore, this subchapter introduces students to the concept of laboratory data analysis. They will learn how to record their observations and measurements, organize data in tables and graphs, and draw conclusions from their findings. Students will also be introduced to basic statistical analysis techniques that can be used to interpret and present their data effectively.

To reinforce the concepts discussed, practical examples and hands-on activities will be provided throughout the subchapter. These activities will allow students to apply their knowledge and skills in a laboratory setting, fostering a deeper understanding of the material covered.

In conclusion, this subchapter serves as a comprehensive introduction to laboratory techniques for students in the field of science education. By emphasizing safety, providing detailed instructions, and incorporating practical examples, students will be well-equipped to navigate the laboratory environment and conduct experiments successfully. Whether pursuing a career in scientific research or simply seeking a better understanding of the elements, this subchapter will serve as a valuable resource for students in their scientific journey.

Common Laboratory Apparatus

In the world of chemistry, laboratory apparatus plays a vital role in conducting experiments and carrying out various scientific procedures. These apparatus are essential tools that help students understand the principles of chemistry and conduct experiments effectively. In this subchapter, we will explore some of the most common laboratory apparatus used in scientific experiments.

One of the most basic yet important apparatus is the beaker. Beakers come in various sizes and are used for holding and mixing liquids. They are made of glass or plastic and have clear markings on the side to measure the volume of the liquid accurately. Beakers are versatile and can be used for a wide range of purposes, such as heating substances, conducting reactions, or simply storing liquids.

Another commonly used apparatus is the test tube. Test tubes are small cylindrical tubes made of glass or plastic. They are used for holding small amounts of substances and conducting experiments on a small scale. Test tubes can be heated directly over a Bunsen burner flame or placed in a test tube holder for safe handling. They are often used to observe reactions, perform qualitative tests, or carry out small-scale reactions.

A crucial apparatus in chemistry laboratories is the pipette. Pipettes are used for accurately measuring and transferring small volumes of liquids. They come in various forms, including graduated pipettes and micropipettes. Graduated pipettes are used for transferring larger volumes of liquids, while micropipettes are used for measuring very small volumes accurately. Pipettes are critical in conducting precise experiments, especially when working with small amounts of reagents or solutions.

For heating substances, laboratories often use a Bunsen burner. A Bunsen burner is a gas burner that produces a flame by mixing gas with air. It is widely used for heating test tubes, beakers, or any substance that needs to be heated in a laboratory. Bunsen burners provide a controlled source of heat and are essential for various experiments involving heating or sterilization.

In addition to these apparatus, there are numerous other tools and equipment used in chemistry laboratories, such as funnels, spatulas, droppers, and stirring rods. Each of these apparatus has its unique purpose and significantly contributes to the success of scientific experiments.

Understanding and utilizing common laboratory apparatus is essential for students studying chemistry. These tools allow students to safely and accurately conduct experiments, make observations, and analyze chemical reactions. By becoming familiar with the various laboratory apparatus, students can enhance their understanding of chemistry and develop the necessary skills to become proficient scientists.

In the next subchapter, we will explore different laboratory safety protocols and precautions that students must adhere to while working in a chemistry laboratory.

Safety Guidelines and Procedures

Chemistry is an exciting and essential field of science that allows us to understand the elements and their interactions. However, it is imperative to prioritize safety when conducting experiments or working in a laboratory setting. This subchapter will guide you through the essential safety guidelines and procedures to ensure your well-being and the well-being of those around you.

1. Laboratory Attire and Personal Protective Equipment (PPE): Always wear appropriate attire and PPE, including a lab coat, safety goggles, and gloves, when working with chemicals or conducting experiments. This will protect you from potential spills, splashes, or harmful fumes.

2. Chemical Handling: Handle all chemicals with care and follow the instructions provided by your teacher or lab supervisor. Avoid touching your face or eyes while working with chemicals, and always wash your hands thoroughly after handling them.

3. Ventilation: Work in a well-ventilated area or under a fume hood when dealing with volatile or toxic substances. This will help minimize exposure to harmful fumes or gases.

4. Fire Safety: Familiarize yourself with the location and proper use of fire extinguishers, fire blankets, and emergency exits in the laboratory. Never use an open flame near flammable substances and be cautious when using heating devices or Bunsen burners.

5. Emergency Procedures: Know the emergency procedures of the laboratory, including how to report accidents or injuries. In case of a

spill or chemical exposure, immediately inform your teacher or lab supervisor and follow their instructions.

6. Equipment Handling: Handle laboratory equipment, such as glassware or sharp instruments, with caution. Check for any damages before use and report any issues to your teacher or lab supervisor.

7. Waste Disposal: Dispose of all waste materials, including chemicals and broken glass, in designated containers. Do not pour chemicals down the sink unless instructed to do so.

8. Personal Hygiene: Avoid eating, drinking, or applying cosmetics in the laboratory. Wash your hands thoroughly before leaving the lab to prevent accidental ingestion or contamination.

Remember, safety should always be your top priority when working in a laboratory. By following these guidelines and procedures, you can ensure a safe and enjoyable learning experience in the field of chemistry.

Chapter 14: Review and Practice

Key Concepts and Definitions

In the study of chemistry, there are several key concepts and definitions that students need to understand in order to grasp the fundamentals of this fascinating field. This subchapter aims to introduce these essential ideas in a clear and concise manner, providing students with a solid foundation to build upon as they delve deeper into the world of chemistry.

One of the fundamental concepts in chemistry is the idea of an element. An element is a pure substance that cannot be broken down into simpler substances by chemical means. It is composed of atoms, which are the basic building blocks of matter. Each element is uniquely identified by its atomic number, which corresponds to the number of protons in its nucleus. The periodic table of elements is a graphical representation that organizes all known elements based on their atomic number and other properties.

Another important concept in chemistry is the notion of a compound. A compound is a substance made up of two or more different elements that are chemically bonded together. These bonds can be either ionic or covalent, depending on the nature of the elements involved. Understanding the properties and behavior of compounds is crucial for studying chemical reactions and their applications in various fields.

Furthermore, it is essential to grasp the concept of a molecule. A molecule is a group of atoms held together by chemical bonds. Molecules can consist of atoms of the same element or different elements. They can be either simple or complex, and their structure and composition play a vital role in determining their properties and reactivity.

In addition to these concepts, this subchapter will also define other important terms such as matter, energy, and chemical reactions. Matter refers to anything that occupies space and has mass, while energy is the capacity to do work or produce heat. Chemical reactions involve the rearrangement of atoms to form new substances, accompanied by the release or absorption of energy.

By familiarizing themselves with these key concepts and definitions, students will gain a solid understanding of the fundamental principles of chemistry. This knowledge will serve as a strong foundation for further exploration into the intricate world of the elements and their interactions.

Practice Problems and Exercises

In the exciting world of chemistry, understanding the elements is crucial. To truly grasp the concepts and principles of this fascinating subject, it is essential to put your knowledge into practice. That's where practice problems and exercises come in. This subchapter is designed to help you reinforce your understanding of the elements through hands-on learning.

Whether you are a high school student, college student, or even a self-learner, this section is for you. By actively engaging with practice problems and exercises, you will deepen your understanding and gain confidence in your knowledge of chemistry.

The practice problems in this subchapter cover a wide range of topics, from the periodic table and atomic structure to chemical reactions and stoichiometry. Each problem is carefully crafted to challenge your thinking and test your comprehension of the material. By working through these problems, you will develop problem-solving skills that are essential for success in chemistry.

Furthermore, the exercises provided in this subchapter are designed to give you a hands-on experience with the elements. From conducting simple experiments to analyzing data, these exercises will help you apply your theoretical knowledge to real-world situations. By actively engaging in these activities, you will gain a deeper appreciation for the practical applications of chemistry.

Remember, practice makes perfect. The more you practice solving problems and performing experiments, the more proficient you will become in chemistry. Don't be afraid to make mistakes along the way;

they are an essential part of the learning process. Through trial and error, you will gain valuable insights and refine your understanding of the elements.

To make the most of this subchapter, we encourage you to work through the practice problems and exercises systematically. Start by reviewing the relevant concepts and principles covered in the preceding chapters. Then, attempt the problems and exercises, allowing yourself enough time to think critically and solve them independently. Finally, check your answers and review any mistakes or areas of weakness.

By engaging with the practice problems and exercises in this subchapter, you will enhance your understanding of the elements and develop the skills necessary to excel in chemistry. So, grab your pencil, put on your lab coat, and get ready to dive into the exciting world of chemistry through active learning.

Answer Key

In this subchapter, we provide you with the answer key to the practice questions and exercises presented throughout the book, "Chemistry Made Easy: A Student's Guide to Understanding the Elements." As students in the field of science education, it is crucial to have a comprehensive understanding of the key concepts and principles of chemistry. The answer key serves as a valuable tool to help you assess your knowledge and gauge your progress.

As you navigate through the chapters of this book, you will encounter a variety of practice questions and exercises designed to reinforce your learning. These questions range from simple multiple-choice questions to more complex problem-solving exercises. The answer key provided in this subchapter will allow you to check your answers and ensure that you have grasped the concepts correctly.

By using the answer key, you can identify any areas where you may need further clarification or additional practice. It is essential to thoroughly review the explanations provided for each correct answer and understand why it is the correct choice. This process of self-assessment and reflection will enhance your understanding of the subject matter and help you identify any gaps in your knowledge.

Additionally, the answer key can be used as a study resource. As you work through the questions and exercises, you can refer to the answer key to verify your answers. This active engagement with the material will reinforce your learning and improve your retention of the information.

Remember, the answer key should be used as a tool for self-assessment and learning, rather than as a shortcut to avoid understanding the material. It is important to attempt the questions independently before referring to the answer key. This way, you can fully challenge yourself and develop problem-solving skills that are essential in the field of science.

In conclusion, the answer key provided in this subchapter is an invaluable resource for students of science education. It allows you to assess your understanding of the concepts and principles covered in this book, and it aids in identifying areas for improvement. Utilize the answer key to reinforce your learning and enhance your knowledge of chemistry.

Chapter 15: Conclusion and Further Resources

Recap of Key Points

In this subchapter, we will recapitulate the essential points covered thus far in our book, "Chemistry Made Easy: A Student's Guide to Understanding the Elements." Understanding these key concepts is crucial for students in the field of science education.

1. Introduction to Chemistry: We began by introducing the subject of chemistry and its importance in our daily lives. Chemistry is the study of matter, its properties, and the changes it undergoes. It helps us understand the structure of matter, reactions, and how substances interact.

2. Atomic Structure: We explored the fundamental unit of matter, the atom. Atoms comprise protons, neutrons, and electrons. Protons and neutrons are located in the nucleus, while electrons orbit around the nucleus in energy levels or shells. We also discussed the periodic table and its organization.

3. Chemical Bonding: Chemical bonding helps us understand how atoms interact and form compounds. We covered the types of chemical bonds, including ionic, covalent, and metallic bonds. We also learned about Lewis dot structures and how to determine the number of valence electrons.

4. Chemical Reactions: Chemical reactions involve the rearrangement of atoms to form new substances. We studied different types of reactions, such as synthesis,

decomposition, combustion, and single and double displacement reactions. Understanding these reactions helps predict products and balance equations.

5.   States   of   Matter:
The three states of matter - solid, liquid, and gas - were explored in detail. We discussed their properties, intermolecular forces, and phase changes. Additionally, we explained concepts related to solutions, solubility, concentration, and factors affecting reaction rates.

6.   Acids   and   Bases:
Acids and bases play a vital role in chemistry. We covered the properties of acids and bases, pH scale, and indicators. We also discussed the concept of neutralization reactions and how to calculate the concentration of solutions using titration.

7.   Electrochemistry:
Electrochemistry deals with the study of chemical reactions involving electricity. We studied redox reactions, half-reactions, oxidation numbers, and electrochemical cells. Understanding these concepts is crucial for applications such as batteries and electrolysis.

By recapping these key points, we hope to reinforce your understanding of the fundamental concepts covered in this book. Remember to review and practice these concepts regularly to master the subject. Chemistry can be challenging but with dedication and perseverance, you can excel in this fascinating field of science.

Additional Resources for Further Study

In the pursuit of knowledge and understanding, it is important for students to have access to a wide range of resources that can supplement their learning. While this book, "Chemistry Made Easy: A Student's Guide to Understanding the Elements," aims to provide a comprehensive introduction to the world of chemistry, there are plenty of additional resources available that can help students delve deeper into this fascinating subject. This subchapter highlights some of the top resources that students can explore for further study in science education.

1.   Online   Educational   Platforms: In today's digital age, online educational platforms have become a valuable tool for students. Websites such as Khan Academy, Coursera, and edX offer a vast array of chemistry courses, lectures, and tutorials. These platforms often provide interactive quizzes and assignments to test and reinforce learning. Students can access these resources from the comfort of their homes, allowing for flexible and self-paced study.

2.   Scientific   Journals   and   Publications: For those seeking more advanced knowledge, scientific journals and publications are an invaluable resource. Peer-reviewed journals like the Journal of Chemical Education and Nature Chemistry publish cutting-edge research articles, reviews, and experiments. By reading these publications, students can gain insight into the latest discoveries and developments in the field of chemistry.

3.   Chemistry   Textbooks: While this book serves as an excellent introductory guide, students can

benefit from exploring additional textbooks. Textbooks like "Chemistry: The Central Science" by Brown, LeMay, and Bursten or "General Chemistry" by Linus Pauling provide in-depth explanations and exercises that can help reinforce concepts and build a strong foundation in chemistry.

4. Chemistry Tutoring Services: Sometimes, students may need personalized guidance to overcome specific challenges in their chemistry studies. Tutoring services, whether in-person or online, can provide one-on-one support, tailored explanations, and additional practice materials. Tutors can help students grasp difficult concepts, solve complex problems, and enhance their overall understanding of the subject.

5. Chemistry Apps and Simulations: In the digital age, there are various chemistry apps and simulations available for students to explore. These interactive tools provide a hands-on approach to learning chemistry, allowing users to visualize molecules, perform virtual experiments, and practice problem-solving. Some popular chemistry apps include "MEL Chemistry," "Chemical Equation Balancer," and "The Elements: A Visual Exploration."

By utilizing these additional resources, students can enhance their understanding of chemistry and further develop their scientific knowledge. Remember, the journey of learning is a continuous one, and these resources are here to support and supplement your growth as a student of science education.

Final Thoughts and Encouragement

Congratulations, dear students, on reaching the end of this comprehensive guide to understanding the elements! You have embarked on a journey to unravel the wonders of chemistry, and your determination and perseverance have brought you here. As we conclude this book, let us reflect on the knowledge gained and the path that lies ahead.

Chemistry is a captivating subject that fuels our curiosity about the world around us. It is a gateway to understanding the composition, properties, and interactions of matter. By studying the elements, you have gained a solid foundation in this fascinating field. But remember, this is just the beginning. The world of chemistry extends far beyond what we have covered here, and there is so much more to explore.

As you continue your scientific journey, always keep in mind the importance of curiosity and critical thinking. These two traits will serve as your guiding lights in unraveling the mysteries of the elements. Never stop asking questions, for it is through questioning that new discoveries are made. Embrace the challenges that come your way, for they are opportunities for growth and learning.

Remember that science education is not just about memorizing facts and formulas, but about understanding the underlying concepts. Seek to grasp the fundamental principles, and you will find that chemistry becomes more intuitive and enjoyable. Don't be discouraged by complex equations or unfamiliar terms. Take small steps, break things down, and gradually build your knowledge. As the saying goes, "Rome wasn't built in a day."

In your pursuit of scientific knowledge, surround yourself with like-minded individuals who share your passion for chemistry. Engage in discussions, collaborate on projects, and learn from each other's experiences. Science is a collective effort, and by joining forces, you can achieve remarkable feats.

Lastly, remember to celebrate your successes, no matter how small they may seem. Each breakthrough, each "aha" moment, is a testament to your hard work and dedication. Embrace the joy of discovery and let it fuel your desire to delve deeper into the world of chemistry.

So, dear students, as you close this book, let it serve as a stepping stone to your future endeavors. The study of the elements is a lifelong pursuit, and you are now equipped with the tools to continue this exploration. Embrace the challenges, savor the victories, and always remain curious. The world of chemistry awaits your discoveries.